Cambridge Primary
Mathematics

Second Edition

Workbook 4

Steph King
Josh Lury

Series editors:
Mike Askew
Paul Broadbent

Boost

HODDER
EDUCATION
AN HACHETTE UK COMPANY

Cambridge International copyright material in this publication is reproduced under licence and remains the intellectual property of Cambridge Assessment International Education.

Third-party websites and resources referred to in this publication have not been endorsed by Cambridge Assessment International Education.

Registered Cambridge International Schools benefit from high-quality programmes, assessments and a wide range of support so that teachers can effectively deliver Cambridge Primary. Visit www.cambridgeinternational.org/primary to find out more.

Acknowledgements

The Publishers would like to thank the following for permission to reproduce copyright material.

Photo credits
p. 8 *tl, cr,* **p. 16** *tl, cr,* **p. 23** *tl, cr,* **p. 28** *tl, cr,* **p. 32** *tl, cr,* **p. 37** *tl, cr,* **p. 43** *tl, cr,* **p. 47** *tl, cr,* **p. 52** *tl, cr,* **p. 57** *tl, cr,* **p. 63** *tl, cr,* **p. 67** *tl, cr,* **p. 73** *tl, cr,* **p. 77** *tl, cr,* **p. 83** *tl, cr,* **p. 87** *tl, cr,* **p. 92** *tl, cr,* **p. 96** *tl, cr* c Stocker Team/Adobe Stock Photo.

t = top, *b* = bottom, *l* = left, *r* = right, *c* = centre

Every effort has been made to trace all copyright holders, but if any have been inadvertently overlooked, the Publishers will be pleased to make the necessary arrangements at the first opportunity.

Hachette UK's policy is to use papers that are natural, renewable and recyclable products and made from wood grown in well-managed forests and other controlled sources. The logging and manufacturing processes are expected to conform to the environmental regulations of the country of origin.

Orders: please contact Hachette UK Distribution, Hely Hutchinson Centre, Milton Road, Didcot, Oxfordshire, OX11 7HH. Telephone: +44 (0)1235 827827. Email education@hachette.co.uk Lines are open from 9 a.m. to 5 p.m., Monday to Friday. You can also order through our website: www.hoddereducation.com

ISBN: 978 1 3983 01207

© Steph King and Josh Lury 2021

First published in 2017

This edition published in 2021 by

Hodder Education,

An Hachette UK Company

Carmelite House

50 Victoria Embankment

London EC4Y 0DZ

www.hoddereducation.com

Impression number   10 9 8 7 6 5 4 3 2 1
Year   2025 2024 2023 2022 2021

Cover illustration by Lisa Hunt, The Bright Agency

Illustrations by Alex van Houwelingen, Ammie Miske

Typeset in FS Albert 15/17 by IO Publishing CC

Printed in the UK

A catalogue record for this title is available from the British Library.

Contents

Unit 1   Number

> Remember: When you see this star ⭐, it is showing you that the activity develops your Thinking and Working Mathematically skills!

Can you remember?

Read these numbers. Then use digits to write them.

a One hundred and twenty-three → ☐ ☐ ☐

b Three hundred and twenty-one → ☐ ☐ ☐

c Four hundred and four → ☐ ☐ ☐

d Four hundred and forty → ☐ ☐ ☐

Negative numbers

1 Fill in the missing numbers.

a

−4				0	1		3	

b

				−2		0		2

2 Complete each sentence using the words in the boxes.
You may use the words more than once.

zero   positive   on a number line   to the left of

to the right of   negative   and is placed

a The number −10 is _____

_____

b The number 10 is _____

_____

Counting on and back

1 For each count below, use at least four of these numbers.
You may use the numbers more than once.

| 6 | 12 | −2 | 0 | 7 | 2 | −3 | 4 | 3 |

a Count back in twos: _____

b Count on in fives: _____

c Count back in threes: _____

2 Read the counting rule. Then fill in each counting pattern.

	Start number						End number	Counting rule
a	125							+5
b		136						+100
c			452					+10
d				214				−10
e					1 047			−100
f						209		−5

Number and place value

1 Fill in each bar model to show the place value parts.

a

1 224			
1 000			4

b

9 497			

c

13 348			
		40	

d			25 052		

2 Write the missing numbers in this count of 1 000.

( ) (2 576) ( ) (4 576) ( )

3 How many? Fill in this table.

	Number	1 000s	100s	10s
a	4 352	4		
b	14 352		143	
c	40 302			

Comparing and ordering numbers

1 Use four or five of these digits to make up correct < or > statements.

6   4   3   4   5

a ( )( )( )( ) < ( )( )( )( )

b ( )( )( )( ) > ( )( )( )( )

c ( )( )( )( )( ) < ( )( )( )( )( )

d ( )( )( )( )( ) > ( )( )( )( )( )

Look at these thermometers.

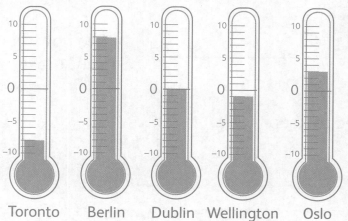

Place	Temperature in degrees (°C)
Toronto	
Berlin	
Dublin	
Wellington	
Oslo	

a Write the correct temperature for each city in the table.

b Use pairs of temperatures to complete each statement.

☐ °C < ☐ °C   ☐ °C > ☐ °C

Rounding numbers

1 Round the numbers under each heading.

Number	Rounded to the nearest 1 000	Rounded to the nearest 100	Rounded to the nearest 10
678			
5 678			
5 245			
9 573			
12 354			

2

I am thinking of a number. I double it.
When I round the answer to the nearest 10 it is 450,
but when I round it to the nearest 100 it is 400.
How many possible starting numbers are there?

Unit 1   Number

Self-check

 I can do this.

 I can do this, but I need to keep trying.

 I can't do this yet.

See how much you know!

What can I do?			
1 I can count back past zero in one-digit steps.			
2 I can count on and count back in steps of one-digit numbers, tens, hundreds and thousands.			
3 I can read and write (symbols and words) whole numbers greater than 1 000.			
4 I can say and explain the value of different digits in a number.			
5 I can decompose numbers to show the place value parts.			
6 I can round numbers to the nearest 10, 100 or 1 000.			
7 I can regroup numbers in different place value parts.			
8 I can place positive and negative numbers around zero on a number line.			
9 I can read and write negative numbers.			
10 I can use negative numbers correctly.			
11 I can compare and order numbers using the symbols =, > and <.			

I need more help with:

_____

_____

_____

Can you remember?

Write the name of each 2D shape.

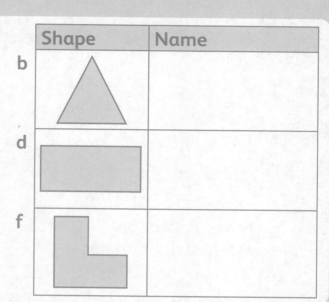

	Shape	Name			Shape	Name
a				b		
c				d		
e				f		

Polygons

 1   Draw six different triangles. Use the dots to guide you.

 2   Now draw four different pentagons and four different hexagons.

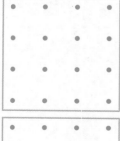 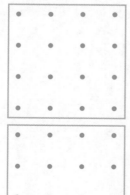 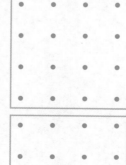 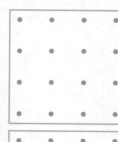

3 Challenge yourself! Follow the instructions for drawing each shape.

a Draw a four-sided shape
with zero right angles.

b Draw a triangle
with one right angle.

c Draw a pentagon
with zero right angles.

d Draw a hexagon with
two right angles.

e Draw a polygon with an odd
number of sides and an even
number of right angles.

f Draw a polygon with an even
number of vertices and an
odd number of right angles.

Compound shapes and tessellation

1 Make a pattern! Draw copies of the shaded shape in the grid. Fill as much space as you can.

a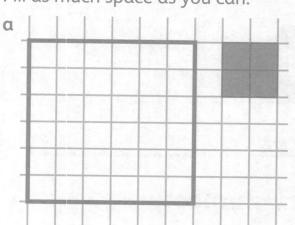

b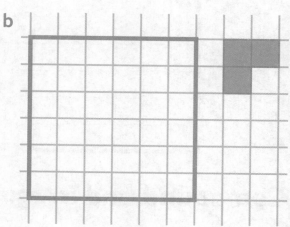

2 Try to fill each grid with the shaded shape.

a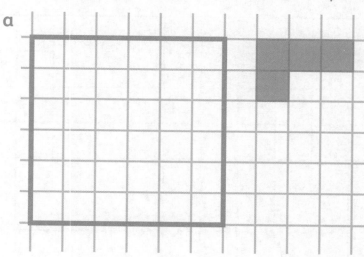

How many squares are left over?

There are _____ squares left over.

b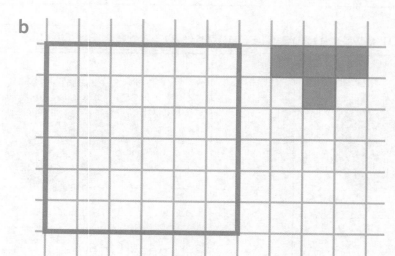

There are _____ squares left over.

3 Copy and join the four shapes to make one compound shape.

a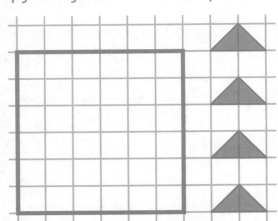

b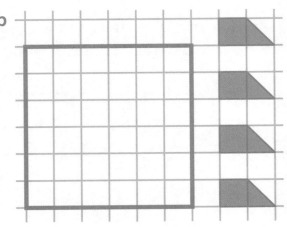

Compound and irregular shapes

1 Sort these shapes in order, from shortest to longest perimeter.

_____ , _____ , _____ , _____ , _____ , _____ , _____

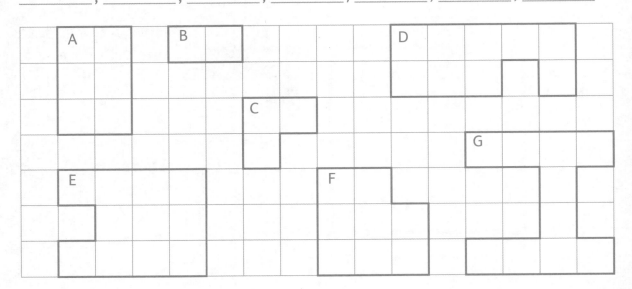

2 Estimate the area of each shape in squares.

Key | 1 square = 1 cm

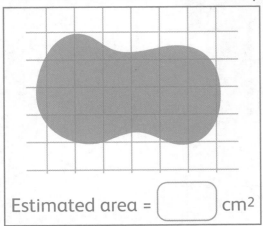

Estimated area = [   ] cm²

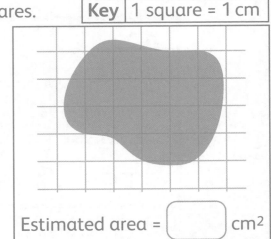

Estimated area = [   ] cm²

3 Choose any three shapes. Add to each to make an area of 10 cm².

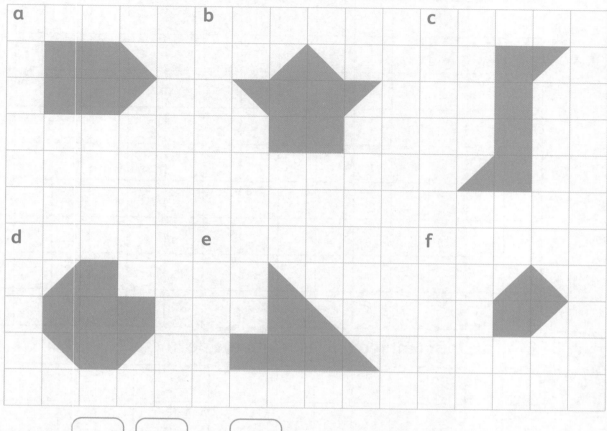

Shapes ⬜ , ⬜ and ⬜ now have an area of 10 cm².

Squares and rectangles

1 Use the measurements in the table below to draw five rectangles.
Use the grid on the next page.
Then calculate the area of each rectangle and write it under **Area**.

	Width	Height	Area
a	3 cm	2 cm	
b	3 cm	3 cm	
c	3 cm	4 cm	
d	30 mm	50 mm	
e	1.5 cm	6 cm	

Key | 1 square = 1 cm

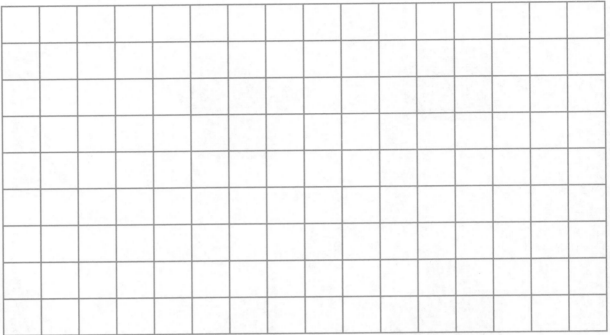

2 Draw six different rectangles. Write the perimeter of each in cm.

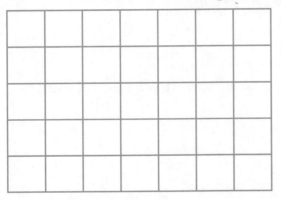

a Perimeter = [ ] cm

b Perimeter = [ ] cm

c Perimeter = [ ] cm

d Perimeter = [ ] cm

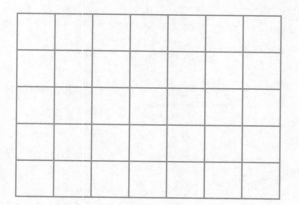

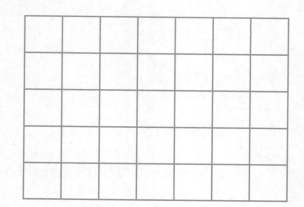

e Perimeter = ☐ cm

f Perimeter = ☐ cm

3 Draw two shapes below. One must have the same **area** as Shape A. The other must have the same **perimeter** as Shape B. See if you can find different solutions.

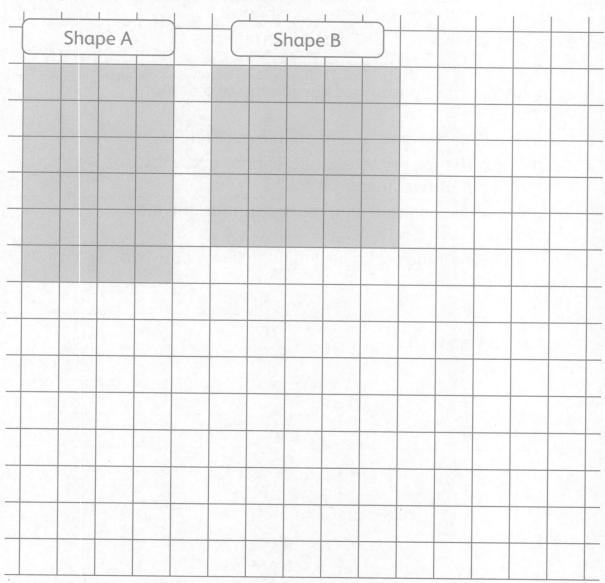

Shape A

Shape B

Self-check

	I can do this.
	I can do this, but I need to keep trying.
	I can't do this yet.

What can I do?	😃	😐	🙁
1 I can recognise, name and describe polygons.			
2 I can describe the properties of 2D shapes.			
3 I can sort 2D shapes based on their properties.			
4 I can show when 2D shapes tessellate.			
5 I can make compound shapes by joining two or more 2D shapes.			
6 I can calculate the area of a rectangle in different units.			
7 I can measure the perimeter of 2D shapes and calculate the perimeter of a rectangle.			
8 I can use whole and part squares to estimate the area of a shape.			
9 I can measure using scales, including halfway between marks on a scale.			

I need more help with:

_____

_____

_____

Unit 3   Calculation

Can you remember?

How many tens?		How many hundreds?	
832 → **83** tens	d	235 → ☐ hundreds	
a 382 → ☐ tens	e	4 235 → ☐ hundreds	
b 1 382 → ☐ tens	f	2 453 → ☐ hundreds	
c 3 218 → ☐ tens	g	12 453 → ☐ hundreds	

Skills for calculating

1 Fill in the numbers to show the value of each digit.

a   353 = ☐ + ☐ + ☐

b   3 535 = ☐ + ☐ + ☐ + ☐

c 35 357 = ☐ + ☐ + ☐ + ☐ + ☐

d   5 057 = ☐ + ☐ + ☐ + ☐

2 Use only the digits **3** and **8** each time to complete these statements.

a ☐☐☐ rounds to 400 to the nearest 100

b ☐☐☐☐ rounds to 3 000 to the nearest 1 000

c ☐☐☐☐ rounds to 3 400 to the nearest 100

d ☐☐☐☐ rounds to 8 340 to the nearest 10

e ☐☐☐☐☐ rounds to 40 000 to the nearest 10 000

3 Regroup each number. You may use the values more than once.

| 450 | 24 | 1 200 | 300 | 75 |

a 1 275 = ☐ + ☐   b 474 = ☐ + ☐

c 1 575 = ☐ + ☐ + ☐

d 1 650 = ☐ + ☐

e 1 725 = ☐ + ☐ + ☐

Using rounding to help with adding and subtracting

1 Write the missing information.

a 78 INPUT

+19 ⊙⊙ OUTPUT ☐

b ☐ INPUT

+31 ⊙⊙ OUTPUT 147

c ☐ INPUT

−98 ⊙⊙ OUTPUT 238

d 347 INPUT

−102 ⊙⊙ OUTPUT ☐

Working with addition

1 Decide if the estimate will be **greater than** or **less than** the answer.

Calculation	> or <	Estimate		Calculation	> or <	Estimate
a 21 + 61	☐	20 + 60	b	29 + 59	☐	30 + 60
121 + 161	☐	120 + 160		229 + 259	☐	230 + 260
c 33 + 73	☐	30 + 70	d	65 + 75	☐	70 + 80
333 + 473	☐	330 + 470		165 + 275	☐	170 + 280

2 Solve each problem. Calculate in your own way.

Problem	My working out
a A jug holds 345 ml of water. Jin adds 255 ml of water. How many millilitres (ml) of water are in the jug in total? Answer = ⬭	
b A book costs 675 cents. A game costs 244 cents more. How much is the game? Answer = ⬭	
c A length of wood is 348 cm. Another piece is 157 cm longer. What is the length of the longer piece of wood? Answer = ⬭	

3 Choose from these digits to make addition calculations. You may use each digit more than once.

⬭ 3   ⬭ 5   ⬭ 0   ⬭ 1   ⬭ 6

a It must have a four-digit answer that is greater than 5 000.

_____

b It must have a three-digit answer that is less than 800.

_____

Working with subtraction

1 Estimate. Then calculate.

Calculation	Estimate	Choose your own method
a 387 – 139		
b 435 – 168		
c 653 – 297		

2 Banko uses regrouping to help him complete this subtraction.

521 – 245 = 521 – 200 – 21 – 23

a Explain the mistake that Banko has made.

_____

_____

b Correct Banko's mistake and find the answer to 521 – 245.

_____

_____

3 Use each digit once to make two numbers with a difference that rounds to 110 to the nearest 10.

0   4   5   6   7   9

[ ][ ][ ] and [ ][ ][ ]

Multiplication and division facts

1 Draw an array first. Then complete each multiplication.

a 2 × 3 = [   ]

b 4 × 3 = [   ]

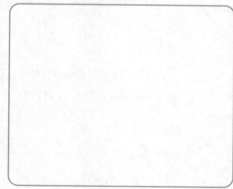

c 8 × 3 = [   ]

d 4 × 6 = [   ]

2 Write a division fact to match each array in question 1.

a _____

b _____

c _____

d _____

3 The pictogram shows the number of visitors at four different museums.

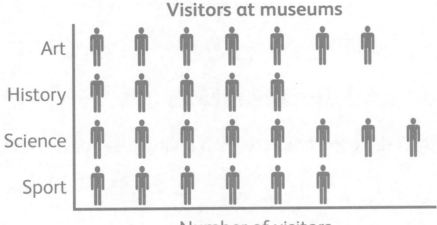

Visitors at museums

Work out how many people visited each museum when 🚹 has the value shown in each row of this table.

	Art	History	Science	Sport
Number of visitors when 🚹 has the value **3**				
Number of visitors when 🚹 has the value **6**				
Number of visitors when 🚹 has the value **9**				

Unit 3   Calculation

Self-check

 I can do this.

 I can do this, but I need to keep trying.

 I can't do this yet.

See how much you know!

What can I do?			
1 I can round numbers to the nearest 10, 100 or 1 000.			
2 I can add or subtract with regrouping.			
3 I can decide whether to calculate mentally or use a column method.			
4 I can use rounding to estimate answers, and say if the actual answer is less than or greater than the estimate.			
5 I can check if an answer is incorrect by estimating.			
6 I can decompose numbers to show the place value parts.			
7 I can regroup numbers in a variety of ways.			
8 I can use decomposing and regrouping to make calculations easier.			
9 I can use the relationships between the 2, 4 and 8 times tables.			
10 I can use the relationships between the 3, 6 and 9 times tables.			
11 I can state related division facts for a given multiplication fact.			
12 I can apply calculating skills and known facts to problems.			

I need more help with:

_____

_____

Can you remember?

Draw hands on the clocks to show these times.

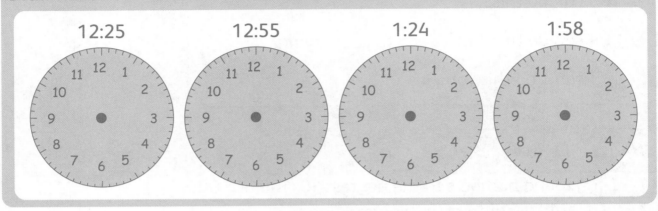

12:25   12:55   1:24   1:58

12- and 24-hour clocks

1 Draw lines to join the matching pairs.

2:15 p.m.	Half-past nine in the evening
9:30 a.m.	Quarter to three in the afternoon
3:15 p.m.	Quarter past two at night
9:30 p.m.	Quarter past two in the afternoon
2:15 a.m.	Quarter past three in the afternoon
2:45 p.m.	Half-past nine in the morning

2 Draw a likely activity for each time.

a

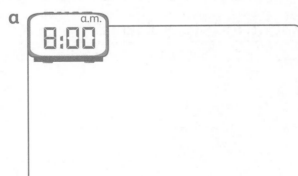

8:00 a.m.

b

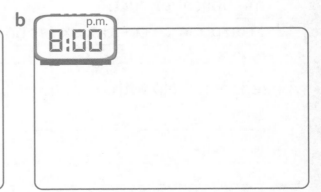

8:00 p.m.

3 Use all four digits each time to make six different 24-hour times. Draw the hands on the clocks.

3   0   5   2

a [ ][ ] : [ ][ ]

b [ ][ ] : [ ][ ]

c [ ][ ] : [ ][ ]

d [ ][ ] : [ ][ ]

e [ ][ ] : [ ][ ]

f [ ][ ] : [ ][ ]

4 Write the equivalent times.

a 1 minute = [ ] seconds   **b** 10 minutes = [ ] seconds

c 5 minutes = [ ] seconds   **d** $\frac{1}{2}$ minute = [ ] seconds

e [ ] minutes = 1 hour   **f** 90 minutes = [ ] hours

g 180 minutes = [ ] hours   **h** [ ] minutes = 10 hours

Calendars and timetables

1 **a** Before you fill in the calendar, predict how many of each day there will be in this month of May.

I predict ☐ Mondays. I predict ☐ Fridays.

I predict ☐ Tuesdays. I predict ☐ Saturdays.

I predict ☐ Wednesdays. I predict ☐ Sundays.

I predict ☐ Thursdays.

May						
Sun	Mon	Tue	Wed	Thurs	Fri	Sat
					25	26
27						

b Now fill in the calendar. Were your predictions in part **a** correct?

2 Write the times using the 24-hour clock.

	Alphaville Station	Betatown Station	Sigma City	Airport
Express train	9:15 a.m. ☐:☐	10 a.m. ☐:☐	Midday ☐:☐	3 p.m. ☐:☐
Local train	Half-past eleven in the morning ☐:☐	Quarter past midday ☐:☐	Twenty to four in the afternoon ☐:☐	Five past eight in the evening ☐:☐

3 Use the timetable in question 2 to answer these.
   a You arrive at Betatown Station at 11:45.
      Which train can you catch?
   b You need to be at the airport by 22:35.
      Which train will you catch from Alphaville Station?

4 Create a timetable for a special activity day at school.
   Decide on a theme, such as art, sport or books.
   Use the 24-hour clock for the times.

Time	Activity
09:00–09:15	Arrive and welcome

Unit 4   Time

Self-check

 I can do this.

 I can do this, but I need to keep trying.

 I can't do this yet.

See how much you know!

What can I do?			
1 I can use a.m. and p.m. to describe time before and after midday.			
2 I can understand the 24-hour clock, and convert between 12-hour and 24-hour times.			
3 I can explain the differences between the 12-hour and 24-hour clocks.			
4 I can convert between seconds, minutes, hours and days.			
5 I can read and use calendars to solve problems.			
6 I can interpret and use timetables accurately, including the 24-hour clock.			

I need more help with:

_____

_____

_____

Can you remember?

Sort the numbers. Write them in the Venn diagram.

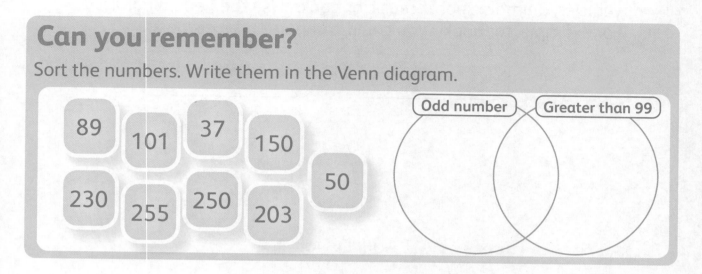

89   101   37   150   230   255   250   203   50

Odd number   Greater than 99

Collecting and sorting data

1   Work with a partner. Sort the list of foods by what you each like and dislike. Then write them in each diagram below.

fish
rice
peas
milk
cabbage
potato
pineapple
carrots
spinach
mango
chicken
soup
salad
soup

Foods I like

Foods _____ likes

	Foods I like	Foods I do not like
Foods _____ likes		
Foods _____ does not like		

2 Collect and compare information on any topic. You could choose vehicles, animals, sports, music or books. Use the Venn diagram. For example, for books, you could compare three authors.

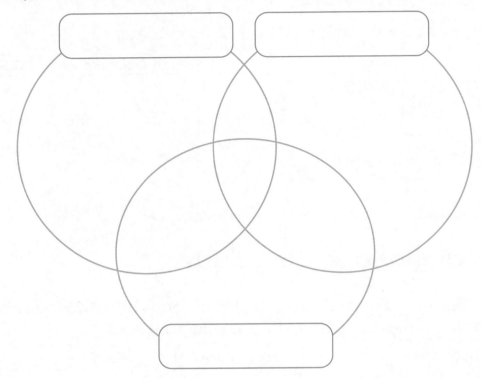

Collecting and comparing information

1 Ask your class how they travel to school. Then fill in this table.

	Tally	Frequency
Walk		
Bicycle		
Car		
Bus		
Other		

2 Here are some tallies from a different class. Fill in the empty column.

	Tally	Frequency
Walk	⊦⊦⊦ ⊦⊦⊦ IIII	
Bicycle	III	
Car	⊦⊦⊦ III	
Bus	⊦⊦⊦ I	
Other	I	

3 Draw bar charts to show your information in questions 1 and 2.

4 **a** Write possible reasons for the different results between the classes.

_____

_____

_____

b What do you think the bar chart would show if you asked 15-year-old learners how they travel to school?

_____

_____

_____

c What would you change about the way people travel to your school?

_____

_____

_____

Unit 5 Statistical methods

Self-check

 I can do this.

 I can do this, but I need to keep trying.

 I can't do this yet.

See how much you know!

What can I do?			
1 I can plan investigations and decide what data to collect.			
2 I can choose methods to collect accurate data.			
3 I can collect data accurately to answer questions, using charts such as tally charts or dot plots.			
4 I can choose how to sort data to show the information clearly.			
5 I can choose how to show data accurately using charts, including the use of different scales.			
6 I can compare data and explain similarities and differences, including answering questions in investigations.			
7 I can interpret data and charts accurately, including understanding how to read scales accurately.			

I need more help with:

_____

_____

_____

Can you remember?

a   12 ÷ 2 = [ ]

b   [ ] ÷ 2 = 5

c   20 ÷ [ ] = 10

16 ÷ 2 = [ ]

[ ] ÷ 2 = 7

8 ÷ [ ] = 4

18 ÷ 2 = [ ]

[ ] ÷ 2 = 9

6 ÷ [ ] = 3

Parts and wholes

1   Complete the bar models.

a
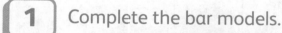
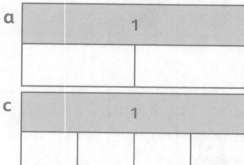

b
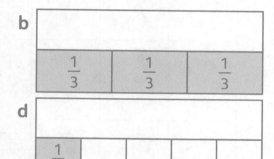

c

d

2   Use the bar models in question 1 to help you answer these.

a   Why is $\frac{1}{3}$ of a whole larger than $\frac{1}{4}$ of the same whole?

_____

b   Why is $\frac{1}{5}$ of a whole smaller than $\frac{1}{2}$ of the same whole?

_____

c   Why are $\frac{4}{4}$ equal to one whole (1)?

_____

Equal shares

1 The teacher divides a whole bag of cherries equally. What fraction of the whole bag will each child get? Write the division sentences you use.

Number of children						Division sentence and fraction
a						☐ ÷ ☐ = ☐/☐
b						☐ ÷ ☐ = ☐/☐
c						☐ ÷ ☐ = ☐/☐
d						☐ ÷ ☐ = ☐/☐
e						☐ ÷ ☐ = ☐/☐

2 Jin divides three 1ℓ cartons of juice equally between four containers. What fraction of a carton of juice is in each container? ☐/☐

Draw something to **convince** your partner that your answer is correct.

Fractions of shapes and quantities

1 Draw straight lines to divide each shape.

a Divide this shape into fifths.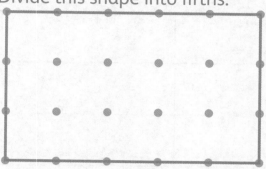

b Divide this shape into thirds.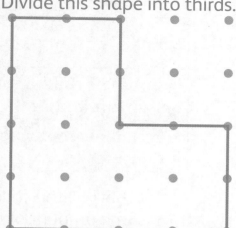

c Divide this shape into tenths.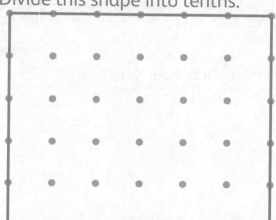

Remember to label the parts of each shape.

2 Draw a circle around the marbles to show each fraction of the whole.

a Show $\frac{1}{4}$ of the group.
How many marbles is $\frac{1}{4}$?
[   ] marbles

b Show $\frac{1}{2}$ of the group.
How many marbles is $\frac{1}{2}$?
[   ] marbles

c Show $\frac{1}{8}$ of the group.
How many marbles is $\frac{1}{8}$?
[   ] marbles

Equivalent fractions

1 The children are sharing the same size fruit pies equally.

a Fill in the table.

Number of fruit pies	1	2		4
Number of children who get an equal share	2		6	8
Fraction of whole pie for each child		$\frac{2}{4}$	$\frac{3}{6}$	

b

I think all the children will get the same amount of a whole pie each!

Do you agree with Pia?

Draw or write something to convince your partner.

2 Another group of children share some vegetable pies equally.

Every pie is the same size. Each child gets the equivalent of $\frac{2}{3}$ of a whole vegetable pie. How many vegetable pies could there be? How many children? Find four solutions.

	Solution 1	Solution 2	Solution 3	Solution 4
Number of vegetable pies				
Number of children who get an equal share				
Fraction of whole pie for each child				

36

Unit 6 Fractions

Self-check

 I can do this.

 I can do this, but I need to keep trying.

 I can't do this yet.

See how much you know!

What can I do?			
1 I can explain why, for example, a length of ribbon cut into four equal parts results in larger pieces than the same size length of ribbon cut into six equal parts.			
2 I can solve simple equal share problems with unit fractions and $\frac{3}{4}$ solutions.			
3 I can use unit fractions to find parts of shapes and quantities.			
4 I can look at a problem and explain why two proper fractions can have an equivalent value.			
5 I can identify simple fractions with a value or total of one and explain why they equal one.			
6 I can apply understanding of fractions to problems.			

I need more help with:

_____

_____

_____

Can you remember?

Write two division facts to go with each multiplication.

a 9 × 3 = 27 _____

b 8 × 7 = 56 _____

_____   _____

Missing number problems

1

I bought one bottle of juice and one sandwich for $5.

I bought the same juice as Sanchia. I bought two bottles for $4.

a What is the cost of one bottle of juice? $_____

b What is the cost of one sandwich? $_____

2 Find the value of the symbols in these missing number problems.

a ☆ − ⬭ = 7

☆ + ☆ = 40

⬭ = _____

☆ = _____

b 24 + ⬠ = ▱

40 − ▱ = 10

⬠ = _____

▱ = _____

c △ + ⬠ = 100

△ + △ = 50

△ = _____

⬠ = _____

Addition and subtraction

1 Play a game with your partner. Choose two numbers from the clouds.
Will you add or subtract? You may choose.
Work out the answer. Then find the number in the grid.
Cover it with a counter.
If your answer is not in the grid, you miss a turn.
The winner is the first person to get three counters in a row.

58   99   342   155   253

497	254	408	97
98	284	311	213
400	441	243	89
187	352	195	157

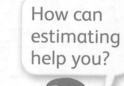

How can estimating help you?

2 Pia has only 25-cent coins in her money box.
She has less than 500 cents in total.
Sanchia has 356 cents in her money box.
Pia has more money than Sanchia.
How much more money could Pia have?
See if you can find all the possible solutions.

Multiplication table of 7

Shade the squares to show that a number of sevens is the total of the same number of twos and fives. Use two colours.
Write the matching multiplication sentences.

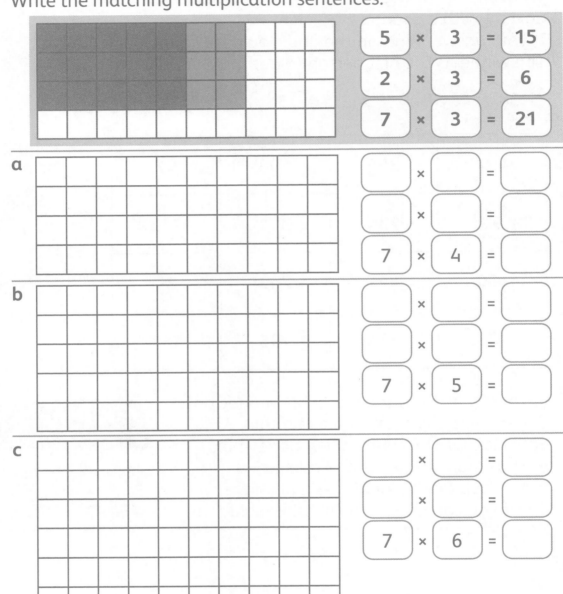

5	×	3	=	15
2	×	3	=	6
7	×	3	=	21

a

	×		=	
	×		=	
7	×	4	=	

b

	×		=	
	×		=	
7	×	5	=	

c

	×		=	
	×		=	
7	×	6	=	

2 Use each number card once to make the division facts true.

| 56 | 7 | 70 | 28 | 8 |

a ⬚ ÷ 7 = 4   **b** ⬚ ÷ 7 = 10

c ⬚ ÷ 7 = ⬚   **d** 49 ÷ ⬚ = 7

Multiplying a 2-digit number by a 1-digit number

 1 Complete these patterns. The first one has been done for you.

2 × 4 = 8

20 × 4 = 80

a   7 × 3 = ☐

70 × 3 = ☐

b   3 × 3 = ☐

30 × 3 = ☐

c   4 × 3 = ☐

40 × 3 = ☐

d   ☐ × 5 = 35

☐ × 5 =350

e   6 × ☐ = 12

6 × ☐ = 120

 2 Solve these. Think about the patterns you notice in the answers.

14 × 7	24 × 7	34 × 7
16 × 7	26 × 7	36 × 7
18 × 7	28 × 7	38 × 7

The patterns I notice are _____

_____

_____

_____

_____

 3 One bottle holds 75 ml.
How many millilitres (ml) in total in nine bottles? _____

Multiplying a 3-digit number by a 1-digit number

1 True or false? Correct any that are false.

		True/False	Correction
a	600 × 4 = 240		
b	7 × 300 = 2 100		
c	500 × 3 = 1 500		
d	3 × 400 = 1 300		

2 Estimate. Then calculate.

	Calculation	Estimate	Use a mental method or a written column method
a	332 × 3		
b	126 × 7		

3 A train journey is 136 km long.
The train repeats the same journey six times.
How many kilometres (km) does the train travel in total? _____

Unit 7   Calculation

Self-check

 I can do this.

 I can do this, but I need to keep trying.

 I can't do this yet.

See how much you know!

What can I do?			
1 I can solve missing number problems and represent them using symbols.			
2 I can decompose numbers to show the place value parts.			
3 I can regroup numbers in a variety of ways.			
4 I can add or subtract with regrouping.			
5 I can decide whether to calculate mentally or use the column method.			
6 I can use rounding to estimate answers, and say if the actual answer will be smaller or larger than the estimate.			
7 I can use decomposing and regrouping to make calculations easier.			
8 I can say when an answer is incorrect by estimating.			
9 I can use known facts to help me complete the seven times table.			
10 I can use the relationships between times tables to help me with calculations.			

I need more help with:

_____

_____

_____

Can you remember?

Write the frequency for each row of this table.

Favourite colour	Tally	Frequency
Red	ⱵⱵ ⱵⱵ ⱵⱵ ⱵⱵ ⱵⱵ ⱵⱵ	
Green	ⱵⱵ ⱵⱵ ⱵⱵ ⱵⱵ I	
Blue	ⱵⱵ ⱵⱵ ⱵⱵ ⱵⱵ ⱵⱵ ⱵⱵ IIII	
Yellow	ⱵⱵ ⱵⱵ ⱵⱵ III	

Certain, impossible, likely, unlikely

1   Write about an event for each probability.
Then draw a picture to show the event.

| Certain | It is certain that _____ |

_____
_____
_____

| Impossible | It is impossible that ___ |

_____
_____
_____

| Likely | It is likely that _____ |

_____
_____
_____

| Unlikely | It is unlikely that _____ |

_____
_____
_____

2 Describe an event each time.
a It is very likely to happen.

_____

_____

b It is equally likely to happen or not happen.

_____

_____

3 Describe an imaginary day full of unlikely events.
Write sentences or draw a comic strip.

I woke up and then ...

And then, at the end of this unlikely day ...

Probability experiments

1 Make a spinner like this.
Complete the sentences. I predict that:

   a On this spinner, the score will

   most often be _____.

   b The score on this spinner is (**more/less**)
likely to be an even number.

Circle 'more' or 'less' in part **b**.

2 Spin the spinner 10 times.
   a Record your results in this table.

Score	Tally	Frequency
1		
3		
5		
6		

   b Are your predictions in question 1 correct at this stage
of the experiment?

3 Spin the spinner another 40 times.
   a Show your results on a dot plot.

   b Are your predictions in question 1 correct now?

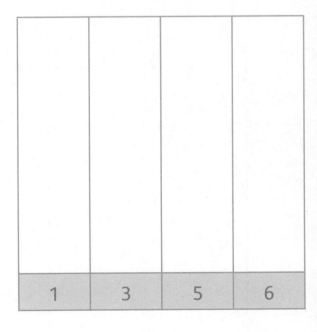

| 1 | 3 | 5 | 6 |

Unit 8   Probability

Self-check

 I can do this.

 I can do this, but I need to keep trying.

 I can't do this yet.

See how much you know!

What can I do?			
1 I can use vocabulary of probability such as *likely*, *unlikely*, *certain* and *impossible* to describe different events.			
2 I can make decisions about the likelihood of an event.			
3 I can complete probability experiments and understand that the number of experiments affects the data.			
4 I can interpret the results of probability experiments to make predictions and decisions.			

I need more help with:

_____

_____

_____

Unit 9   Number

Can you remember?

Draw lines to match numbers with the same values.

 −5   0   −3   5   3

| positive five | negative five | positive three | zero | negative three |

Larger numbers

1 Decompose each number to show its place value parts.

a   343 201 = _____

b   542 097 = _____

c 2 343 201 = _____

2 Circle the numbers that have 345 thousands.

3 450 999   3 452   345 231   354 453   2 345 250

3 Round the numbers to the nearest value under each heading.

Round to the nearest:	100 000	10 000	1 000	100
a 439 505				
b 674 791				
c 1 573 234				

Working with sequences

1 Write the missing numbers in these sequences. Find the rule.

a	110	130	150					Rule: **Add 20**
b	9	5	1					Rule: _____
c		10	20	40			320	Rule: _____
d		96	48		12			Rule: _____

2 Write the value of the next three terms in each sequence.

	Term-to-term rule	First term	Next three terms		
a	Double and add 3	5			
b	Halve and subtract 2	28			
c	Subtract 2 and multiply by 10	3			

3 Which number is not a term in any of the sequences below?

(12)   (19)   (10)   (6)   (16)   (24)   (14)

	Rule	**1st term**
a	multiply by 2	3
b	subtract 5	29
c	add 4	2

The number [ ] is not a term in any of the sequences.

Even and odd numbers

1 Draw counters to show if the totals are odd or even.
Look at the example.

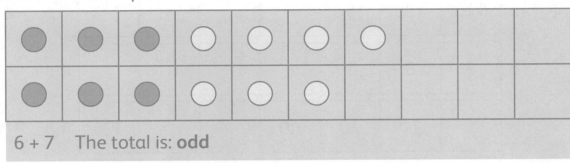

6 + 7   The total is: **odd**

a

4 + 8   The total is: _____

b

5 + 7   The total is: _____

2 Look at the additions and subtractions in the rows and columns of this grid. Write the numbers 2, 5, 7 and 9 in the white squares to complete the grid.

	+		=	even
−		−		−
	+	odd	=	3
=		=		=
odd	+	even	=	

The relationship between factors and multiples

1 **a** Write numbers on the bars to show all the different factors of 18.

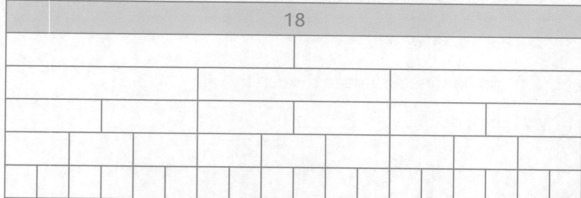

18

b Now write all the factor pairs of 18.

☐ and ☐   ☐ and ☐   ☐ and ☐

2 Write the numbers under the correct headings in this table.
The numbers may appear under more than one heading.

Multiple of 3	Multiple of 5	Factor of 100	Not a multiple of 4

25   30
24   10
12   15

3

4 and 5 are factors of 21 because 4 × 5 = 21.

Explain Banko's mistake.
What should he say?

_____

_____

_____

4 Arrange the digits 1 to 9 in the circles to make a factor pair for each number in the triangular sections.

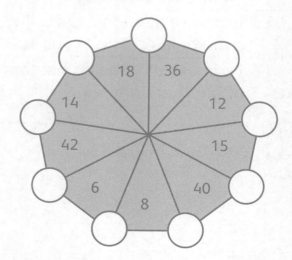

18   36
14   12
42   15
6   40
8

Unit 9　Number

Self-check

 I can do this.

 I can do this, but I need to keep trying.

 I can't do this yet.

See how much you know!

What can I do?			
1 I can decompose numbers to show the place values.			
2 I can explain the values of each digit, given a numeral with repeated digits.			
3 I can compare and order numbers using =, > and <.			
4 I can count on or back in thousands.			
5 I can round numbers to the nearest 10, 100, 1 000 and 100 000.			
6 I can identify and explain the rule for a sequence.			
7 I can continue the sequence if given a rule.			
8 I can count on and back across zero to work with sequences involving positive and negative numbers.			
9 I can explain patterns when adding and subtracting even and odd numbers.			
10 I can find and justify factor pairs, such as 3 and 4 are factors of 12 because 3 × 4 = 12.			
11 I can use division facts to justify multiples, such as 12 is a multiple of 3 and 4, as 12 ÷ 3 = 4 and 12 ÷ 4 = 3.			
12 I can explain when a number is not a factor or multiple of another number.			

I need more help with:

_____

_____

Can you remember?

Estimate the area of this shape. Then measure it to check.

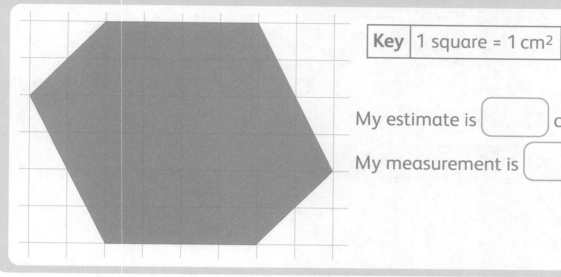

Key | 1 square = 1 cm²

My estimate is ⬚ cm².

My measurement is ⬚ cm².

Symmetry

1 Draw one line of symmetry on each shape.

a

b

c

d

2 Some of these shapes have more than one line of symmetry.
One shape has none. Draw all the lines of symmetry.

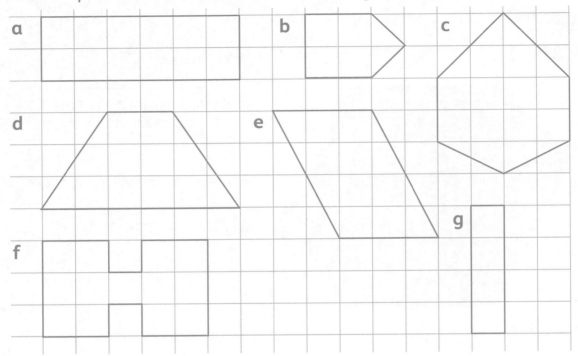

3 Complete these symmetrical patterns.

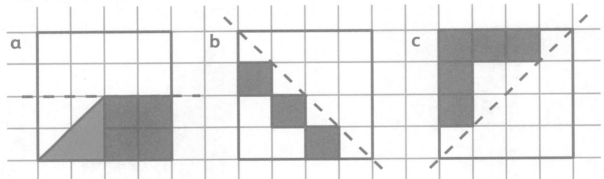

4 Complete these symmetrical shapes. The dotted line is a mirror line.

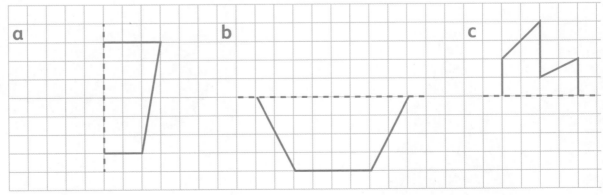

3D shapes and nets

1 Draw the missing face for each net.

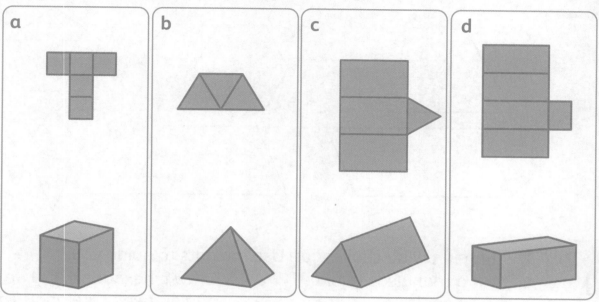

a   b   c   d

2 Complete these prism drawings.
Write the number of vertices, edges and faces.

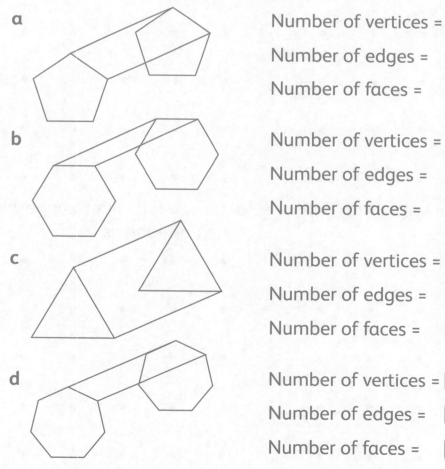

a

Number of vertices = [   ]

Number of edges = [   ]

Number of faces = [   ]

b

Number of vertices = [   ]

Number of edges = [   ]

Number of faces = [   ]

c

Number of vertices = [   ]

Number of edges = [   ]

Number of faces = [   ]

d

Number of vertices = [   ]

Number of edges = [   ]

Number of faces = [   ]

Angles and turns

1 Mark all the acute angles with an **A**, and all the obtuse angles with an **O**.

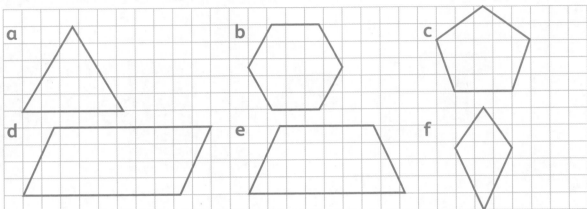

2 Try these drawing challenges. Use the dots to guide you.

a Draw a triangle with one obtuse angle.

b Draw a hexagon with two right angles, two acute angles and two obtuse angles.

c Draw a pentagon with one acute angle.

d Draw a four-sided shape with no right angles.

Self-check

 I can do this.

 I can do this, but I need to keep trying.

 I can't do this yet.

See how much you know!

What can I do?	😃	😐	😟
1 I can name the 2D faces of 3D shapes, and describe their properties.			
2 I can match nets to their matching 3D shapes.			
3 I can find and describe horizontal, vertical and diagonal lines of symmetry on 2D shapes and in patterns.			
4 I can estimate, compare and classify angles, using language such as *right angle*, *acute* and *obtuse*.			
5 I can identify and work with turns, using the vocabulary of *quarter turn*, *half turn*, *three-quarter turn* and *full turn*.			
6 I can connect turns to 90, 180, 270 and 360 degrees.			

I need more help with:

_____

_____

_____

Can you remember?

Tick (✓) the shape that has the largest fraction of the whole shaded.

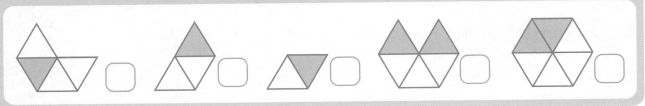

Equal parts

1 Pia and Jin each have an identical bottle of juice.

> I'm going to pour my bottle of juice equally into three glasses.

> I'm going to pour my bottle of juice equally into four glasses.

Whose glasses contain a larger fraction of the bottle? _____
Explain or draw a picture to show how you know.

2 Write a division sentence to match each problem.
Answer each problem as a fraction.

a A baker shares a bag of flour equally between four bowls.
What fraction of the bag is in each bowl?
Division sentence: _____ Answer:

b A group of children arrange themselves into five equal rows.
What fraction of the whole group is in each row?
Division sentence: _____ Answer:

c Sanchia has three slices of toast.
She cuts the toast to share it equally among four friends.
What fraction of a slice of toast does each friend get?
Division sentence: _____ Answer:

Finding fractions of shapes and quantities

1 Circle $\frac{1}{6}$ of the stars.

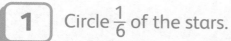

2

I shaded exactly five squares, so I have shaded $\frac{1}{5}$ of this shape.

Explain why Banko is wrong. Banko is wrong because _____

_____

_____

_____.

Shade the correct number of squares to show $\frac{1}{5}$.

3 A group of adults and children are watching a match. There are 36 people in total.

$\frac{1}{4}$ of the group are girls. $\frac{1}{6}$ of the group are boys.

The rest of the group are adults.

How many adults are there? _____

More about equivalent fractions

1 Circle the shape that is **not** $\frac{2}{3}$ shaded.

  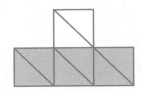 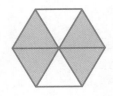

2 Make four shapes, each with a different total number of squares.

Use one colour of pencil crayon to shade $\frac{3}{5}$ of each shape.

Use another colour to complete each whole shape.

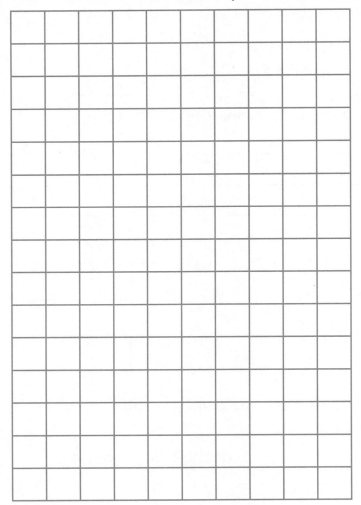

3 Sanchia has a bag of marbles. $\frac{3}{4}$ of them are blue.
There are 15 blue marbles.
What is the total number of marbles in the bag? [   ] marbles

4 Put the fraction cards in the correct places on the number line.

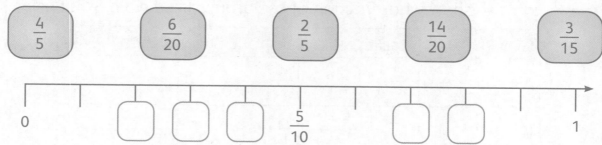

$\frac{4}{5}$ $\frac{6}{20}$ $\frac{2}{5}$ $\frac{14}{20}$ $\frac{3}{15}$

0    $\frac{5}{10}$ 1

Adding and subtracting fractions

1 Use the diagrams to complete each calculation.

a $\frac{1}{5} + \boxed{} = 1$

b $\frac{2}{5} + \boxed{} = 1$

c $\boxed{} + \frac{2}{6} = 1$

d $\boxed{} + \frac{2}{7} = 1$

2 Look at the example. Use two colours of pencil crayons to show each calculation on the diagram.
Fill in the answers.

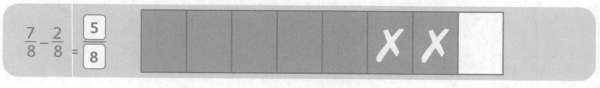

$\frac{7}{8} - \frac{2}{8} = \boxed{\frac{5}{8}}$

a $\frac{1}{5} + \frac{3}{5} = \frac{\boxed{}}{\boxed{}}$

b $\frac{9}{10} - \frac{6}{10} = \frac{\boxed{}}{\boxed{}}$

c $\frac{1}{4} + \frac{2}{4} + \frac{3}{4} = \frac{\boxed{}}{\boxed{}}$

3 Pia and Jin asked groups of people about their favourite fruit. Fill in the table to show what fraction of the groups liked each fruit best.

Group	Apple	Banana	Other
A	$\frac{3}{8}$	$\frac{4}{8}$	
B		$\frac{3}{10}$	$\frac{2}{10}$
C	$\frac{1}{6}$		$\frac{3}{6}$
D	$\frac{3}{9}$	$\frac{4}{9}$	

4 Sanchia has a packet of sports stickers.

$\frac{2}{\Box}$ of the packet are football stickers.

$\frac{3}{\Box}$ of the packet are swimming stickers.

The rest are basketball stickers.
What fraction of the packet could each sport represent?
Find at least three solutions. Record them in your own way.

My solutions

Unit 11 Fractions

Self-check

See how much you know!

 I can do this.

 I can do this, but I need to keep trying.

 I can't do this yet.

What can I do?	😃	😐	🙁
1 I can explain why, for example, a length of ribbon cut into four equal parts results in larger pieces than the same size length of ribbon cut into six equal parts.			
2 I can solve simple equal share problems with unit fractions and $\frac{3}{4}$ solutions.			
3 I can use unit fractions to find parts of shapes and quantities.			
4 In different problems and representations, I can explain why two proper fractions can have an equivalent value.			
5 I can apply understanding of fractions to problems.			
6 I can identify simple fractions with a value or total of one and explain why they equal one.			
7 I can apply addition and subtraction of fractions to problems.			
8 I can estimate, add and subtract fractions with the same denominator.			

I need more help with:

_____

_____

_____

Can you remember?

Draw a compass.
Mark the points north, east, south and west in the correct positions.

North, south, east, west

1 Start on **A** each time. Follow the directions. On which letter do you finish?

a south, west, north, east

I finish on letter ⬚ .

b west, south, east, north

I finish on letter ⬚ .

c west, north-east, south-east, south, west

I finish on letter ⬚ .

d north-west, south-west, south, east, north-west

I finish on letter ⬚ .

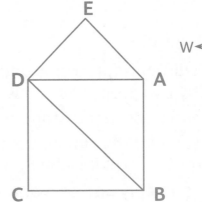

2 Start on A. Finish on B. Visit every letter. Write the instructions.

A→

→B

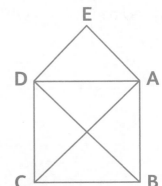

Directions and maps

Use this map of New Zealand to answer the questions below.

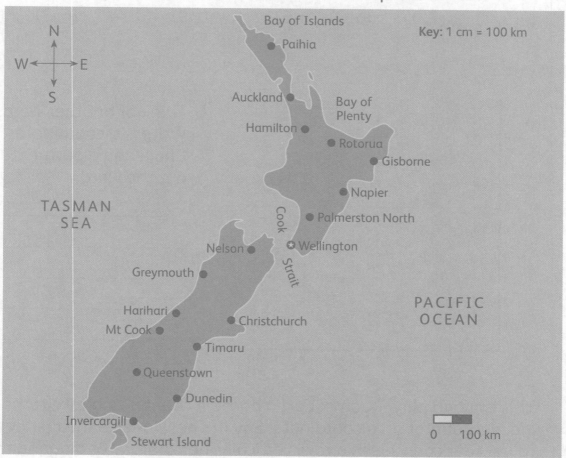

1   Write compass directions to complete these sentences.

a The Tasman Sea is to the _____ of New Zealand.

b The _____ is to the east of New Zealand.

c Stewart Island is in the _____.

2   Find some examples to complete these sentences.

a _____ is further north than _____ .

b _____ is to the south-east of _____ .

c _____ is to the north-west of _____ .

3   Use the scale on the map to complete these.

a It is _____ km from Auckland to Christchurch.

b It is _____ km from Hamilton to Gisborne.

c It is approximately 400 km from Mt. Cook to _____

Coordinates

1 **a** Make a mathematical word! Write the letters of the coordinates.

(5, 7)   (7, 1)   (6, 3)   (6, 5)   (10, 9)   (3, 4)   (6, 3)   (8, 5)

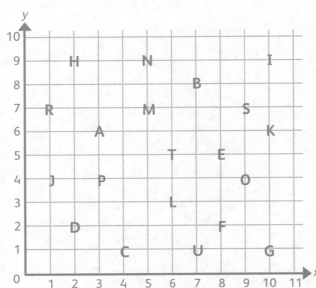

b Think of another word. Write the coordinates. Challenge a partner to work out the word.

_____

_____

_____

_____

_____

2 This is a two-player game. Each choose a counter and take turns. Place a counter on a coordinate. Say it. The winner is the first person to get four counters in a row, vertically, horizontally or diagonally. If you say the wrong coordinate, remove your counter.

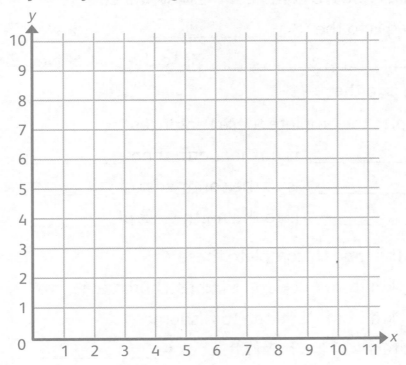

Unit 12 Angles, position and direction

Self-check

 I can do this.

 I can do this, but I need to keep trying.

 I can't do this yet.

See how much you know!

What can I do?			
1 I can identify north, east, south and west on a compass, and understand their position and direction.			
2 I can identify north-east, south-east, north-west and south-west on a compass, and understand their position and direction.			
3 I can use compass directions to give the directions through a maze or map.			
4 I can read a map with a scale and compass directions.			
5 I can interpret a map to answer questions about distance and direction.			
6 I can read and find coordinates on a grid, knowing that coordinates are given in the order: *x*-coordinate, *y*-coordinate.			

I need more help with:

_____

_____

_____

Unit 13 Number

Can you remember?
Write the missing numbers on this number line.

Identifying and building sequences

1 Here are some sequences of numbers.
Write the rule for each. Write the next term.

a (70), (140), (280), ( ) Rule is: _____

b (25), (16), (7), ( ) Rule is: _____

c (1 000), (100), (10), ( ) Rule is: _____

2 Write the missing values in these sequences.

a [1], [4], [ ], [22], [46], [ ], [ ]

Rule is: Multiply by 2 and add 2

b [ ], [66], [30], [ ], [ ]

Rule is: Divide by 2 and subtract 3

c [2], [ ], [14], [ ], [ ]

Rule is: Multiply by 3 and subtract 1

3 Design your own number sequence each time.

a The pattern is in the heading row.

odd	even	odd	even	odd	even	Rule:
◯	◯	◯	◯	◯	◯	

b It has a doubling pattern and includes the term 44.

◯	◯	◯	◯	◯	◯	Rule:

c It involves subtraction and includes the term −4, but not the term 0.

◯	◯	◯	◯	◯	◯	Rule:

d It involves addition and multiplication, and includes the term 10.

◯	◯	◯	◯	◯	◯	Rule:

Square numbers

1 **a** Draw counters to show the first four square numbers.

b Now write the matching multiplication facts for the first four square numbers you drew.

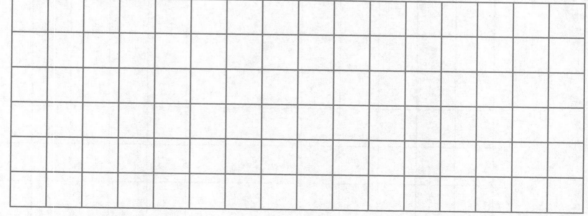

69

a Draw lines to join the square numbers with the arrays.

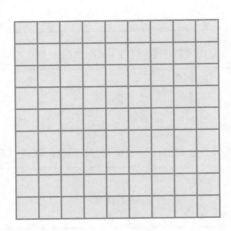 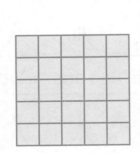 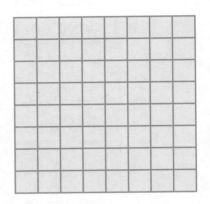

b One array is missing. Write the number of rows and columns it has.

_____

_____

Factors and multiples

a Sketch arrays to show all the factor pairs for 12.

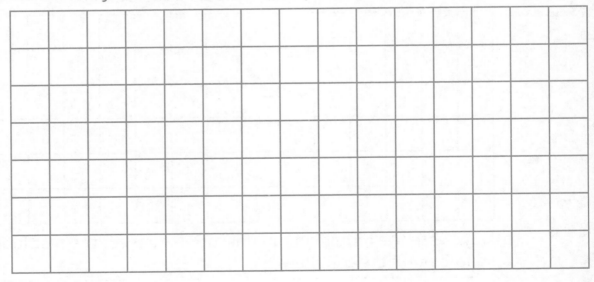

b Write the factor pairs.

_____

2 Answer this question for each number below.
Is the number a factor of 24 or not? Use these words to help you explain your decision. You may use each word more than once.

multiple   divide   factor   exactly

a The number 2 _____

_____

b The number 5 _____

_____

c The number 3 _____

_____

d The number 6 _____

_____

e The number 7 _____

_____

Tests of divisibility

1 Use the numbers on the number cards to complete the sentences below. You may use each number more than once.

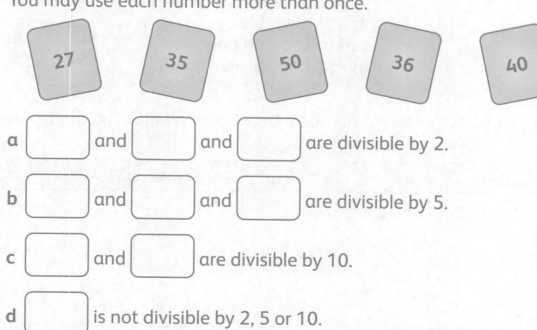

27   35   50   36   40

a ☐ and ☐ and ☐ are divisible by 2.

b ☐ and ☐ and ☐ are divisible by 5.

c ☐ and ☐ are divisible by 10.

d ☐ is not divisible by 2, 5 or 10.

2 Arrange five numbers in the diagram so that:
- the numbers in the four corners total a number that is divisible by 50 but is not divisible by 100
- the numbers on each diagonal total a number that is divisible by 25 but is not divisible by 50.

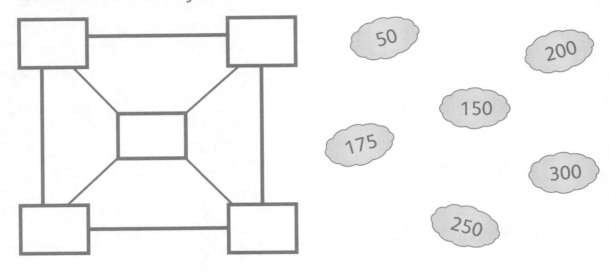

I am thinking of a number that is divisible by 25.
I add 25 and now my number is divisible by 100.

I am thinking of a number that is divisible by 5. I subtract 5 and now my number is divisible by 50.

Think of five different numbers that Guss and Elok could have started with. Show your working here.

Numbers for Guss	Numbers for Elok

Self-check

See how much you know!

 I can do this.

 I can do this, but I need to keep trying.

 I can't do this yet.

What can I do?			
1 I can identify and explain the rule for a sequence.			
2 I can build sequences from a given rule.			
3 I can explain how to build a square number from scratch and how to build the next square number from the previous number.			
4 I can find and justify factor pairs, for example, 3 and 4 are factors of 12 because $3 \times 4 = 12$.			
5 I can use division facts to justify multiples, for example, 12 is a multiple of 3 and 4 because $12 \div 3 = 4$ and $12 \div 4 = 3$.			
6 I can explain when a number is not a factor or multiple of another number.			
7 I can use the tests of divisibility to decide if a number is divisible by 2, 5 or 10.			
8 I can explain why the tests of divisibility for 2, 5 and 10 work.			
9 I can explain why a multiple of 100 must also be a multiple of 25 and 50.			

I need more help with:

_____

_____

_____

Unit 14 Statistical methods

Can you remember?

Add a number to each section of this Venn diagram.

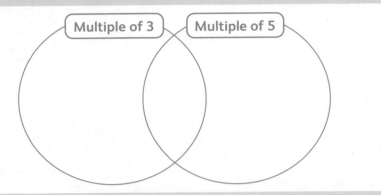

Multiple of 3   Multiple of 5

Interpreting and comparing data

1 Study the information below about four football teams. Then use the information to follow the instructions and answer the questions.

Football team	Frequency of goals scored
United	35
Town	48
Rovers	20
City	15

Matches won	Tally	Frequency
United	⊞⊞ ⊞⊞ ⊞⊞ ⊞⊞	
Town	⊞⊞ ⊞⊞ ⊞⊞ III	
Rovers	⊞⊞ ⊞⊞	
City	IIII	

Complete this bar chart for one of the sets of data above.

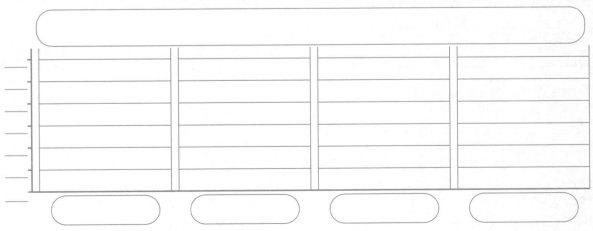

2 Which team do you think is best? Explain your reasons.

_____

_____

_____

3 Which team do you think is worst? Explain your reasons.

_____

_____

_____

Investigating and collecting data

1 Class A learners voted for what they each think is most important to help the environment. The table shows the results.

Class A's vote on the most important environmental issue

Environmental issue	Number of votes
Climate change	ⅢⅢ ⅢⅢ ‖
Reducing pollution	ⅢⅢ ‖‖‖
Animal welfare	ⅢⅢ ⅢⅢ ‖‖‖
Protecting forests	‖‖

Collect information about the issues your class thinks are most important. Ask classmates to vote for one issue each. Choose from:

Climate change   Reducing pollution
Animal welfare   Protecting forests   Other

Use this space to record the information you collect.

75

2 Now decide how to present the information clearly.
Create a chart below.

3 What does your information tell you, compared to the information
that you have from Class A?

_____

_____

_____

_____

_____

4 Describe the information you would need to collect next,
to take this idea further.

_____

_____

_____

_____

_____

Unit 14 Statistical methods

Self-check

 I can do this.

 I can do this, but I need to keep trying.

 I can't do this yet.

See how much you know!

What can I do?			
1 I can plan investigations and decide what data to collect.			
2 I can choose methods to collect accurate data.			
3 I can collect data accurately to answer questions, using charts such as bar charts, tally charts or dot plots.			
4 I can choose how to sort data to show the information clearly.			
5 I can choose how to show data accurately using charts, including the use of different scales.			
6 I can compare data and explain similarities and differences, including using the data to answer questions in investigations.			
7 I can interpret data and charts accurately, including understanding how to read scales accurately.			

I need more help with:

_____

_____

_____

Unit 15 Calculation

Can you remember?

Complete the multiplication triangle. Then use it to help you write four facts.

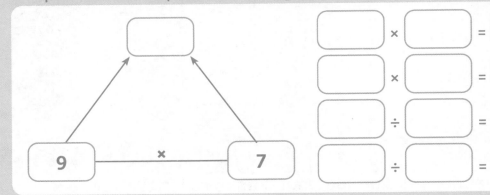

More missing number problems

1 Guss and Elok buy a book and a game for a total of $17.
Guss gets $12 change when he pays for the book with a $20 note.
Use the symbols ⬭ and △ to represent the cost of each item.
Write two number sentences to match the problem.
Use your number sentences to find the cost of the book and the game.

_____ Book = $ _____

_____ Game = $ _____

2 Find the value of one rectangle and the triangle.

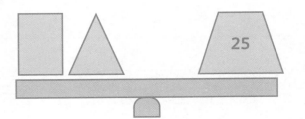
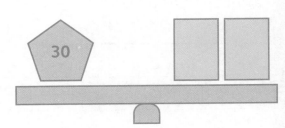

More addition and subtraction

 Look at the puzzles. The numbers in the shaded squares are the difference between the two numbers in the corners. For example, in part **a**, 378 − 137 = 241. Fill in the missing numbers.

a

378	133	
241		
137	105	242

b

423	38	
	99	
		159

 Use all six digit cards to make two 3-digit numbers each time. Before you do this, read the instructions in parts **a** and **b**.

a Make up an addition calculation to give a 4-digit total that is less than 1 500.

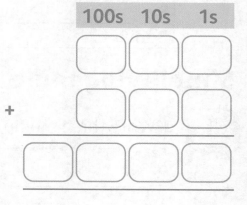

b Make up a subtraction calculation to give an answer that rounds to 300 to the nearest 100.

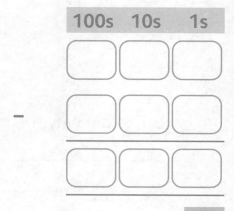

Multiplying and dividing whole numbers by 10 and 100

1 The mass of an African elephant is 100 times heavier than a Sun bear. The mass of a Polar bear is 10 times lighter than that of the African elephant. Write the mass of the African elephant and the Polar bear.

African elephant	Polar bear	Sun bear
[ ] kg	[ ] kg	48 kg

2 The children are saving their money. After one month, they have 10 times as much money as when they started. By the end of the year, they have 100 times as much money as when they started.
Fill in this table to show how each child's money grew.

Name	Starting amount	After a month	After a year
Guss	28 cents		
Elok		310 cents	
Sanchia			2 900 cents
Jin	41 cents		
Pia		460 cents	
Banko			3 400 cents

Simplifying multiplications

1 a Write three multiplication calculations to match this cuboid.

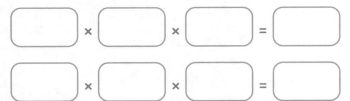

b Explain why the answers in part **a** are the same each time.

_____

_____

2 True or false? Correct any that are false.

	True/False	Correction
18 × 4 = 18 × 2 × 2		
18 × 4 = 9 × 2 × 4		
21 × 3 = 6 × 3 × 3		
7 × 16 = 7 × 7 × 2		

Multiplying larger numbers

1 Choose a number from the rectangle and a number from the hoop. Use what you know about multiples of 10 and 100 to find nine different products. Write them in the table. Look at the example first.

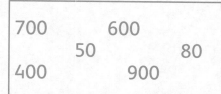

700   600
    50   80
400   900

7
3   4
6   8
5

600 × 7 = 100 × 6 × 7 = 4 200	

2 Sort the calculations into the correct sections of this Venn diagram.

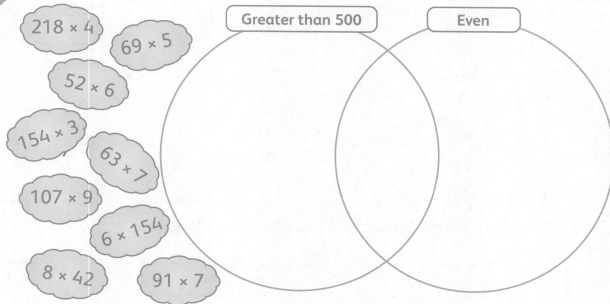

218 × 4

69 × 5

52 × 6

154 × 3

63 × 7

107 × 9

6 × 154

8 × 42

91 × 7

Greater than 500   Even

3 Find the two multiplications in question 2 with the largest product.
Choose your own method to complete each multiplication.

Dividing 2-digit numbers

1 **a** The total cost of six tickets for a play is $96.
What is the cost of one ticket?

My working:_____ $_____

b A length of rope is 56 metres. Jin cuts it into four equal pieces.
What is the length of one piece?

My working:_____ _____metres

2 Work out the remainder each time.

a $61 \div 3 =$ ⬜ remainder ⬜

b $61 \div 4 =$ ⬜ remainder ⬜

c $61 \div 5 =$ ⬜ remainder ⬜

d $61 \div 6 =$ ⬜ remainder ⬜

What do you notice? _____

Does this pattern carry on? Explain your answer.

_____

_____

Unit 15 Calculation

Self-check

 I can do this.

 I can do this, but I need to keep trying.

 I can't do this yet.

See how much you know!

What can I do?			
1 I can solve and represent missing number problems.			
2 I can add or subtract with regrouping.			
3 I can choose to use mental or written methods.			
4 I can use decomposing and regrouping to make calculations easier.			
5 I can use rounding to estimate and check answers.			
6 I can use the relationships between times tables.			
7 I can recall and use multiplication and division facts.			
8 I can use both the associative property and commutative property of multiplication to simplify calculations.			
9 I can multiply and divide whole numbers by 10 or 100 and explain the result using the language of place value.			
10 I can multiply a multiple of 10 or 100 by a single digit.			
11 I can estimate the answer to and multiply a two-digit number by a one-digit number.			
12 I can estimate the answer to and multiply a three-digit number by a one-digit number.			
13 I can estimate and divide whole numbers up to 100 by one-digit whole numbers.			
14 I can apply my skills of calculating and known facts to problems.			

I need more help with:

_____

_____

Can you remember?

Write the time under each clock. Use **a.m.** or **p.m.**

a

b

c

d

_____   _____   _____   _____

Duration

1 Draw hands on the clock faces to show each runner's **End time.**

Runner	Start time	Duration	End time
A		45 minutes	
B		30 minutes	
C		60 minutes	
D		30 minutes	

2 Fill in the duration of each lesson.

Lesson	Start time	End time	Duration
Art	13:30	14:30	
Music	15:05	15:50	
Sport	16:00	18:00	
Dance	19:30	21:00	

Days, weeks, months and years

1 Add these events to the calendar. Write them or add a key with symbols.
  a Banko's party is on the last Saturday of the month.
  b Jin's party is on the first Tuesday of the month.
  c The school play is two weeks after 4 September.
  d The disco is exactly one week before 2 October.
  e The school holiday starts on 29 August and lasts for five days.

September						
Sun	Mon	Tues	Wed	Thurs	Fri	Sat
			14	15	16	17
18	19	20				

Key	

2 Write the date and time of arrival for each journey.

a Depart: 09:00 on 25 November.
Duration of journey: 2 days

Date: _____ Arrive:

b Depart: 13:00 on 29 November.
Duration of journey: 1 day and 10 hours

Date: _____ Arrive:

c Depart: 14:15 on 30 November.
Duration of journey: 1 day and $4\frac{1}{2}$ hours

Date: _____ Arrive:

d Depart: 11:00 on 3 December.
Duration of journey: 1 week and 2 hours

Date: _____ Arrive:

3 Choose an important date in the future. Show how to calculate the number of days, weeks and months until that date.

4 Convert between these durations.

a 24 hours = [   ] days

b [   ] hours = 2 days

c 12 hours = [   ] day

d 2 weeks = [   ] days

e 4 weeks = [   ] days

f 10 weeks = [   ] days

g [   ] weeks = 77 days

Unit 16 Time

Self-check

 I can do this.

 I can do this, but I need to keep trying.

 I can't do this yet.

See how much you know!

What can I do?			
1 I can use a.m. and p.m. to describe time before and after midday.			
2 I can read the 24-hour clock, and convert between 12-hour and 24-hour times.			
3 I can explain the differences between the 12-hour and 24-hour clocks.			
4 I can convert between seconds, minutes, hours and days.			
5 I can read and use calendars to solve problems.			
6 I can interpret and use timetables accurately, including the 24-hour clock.			
7 I can interpret the duration of an event as the length of time it takes, from start to finish/beginning to end.			
8 I can calculate the duration of an event within one day.			
9 I can calculate the duration of an event that takes longer than one day.			

I need more help with:

_____

_____

_____

Unit 17 Fractions and percentages

Can you remember?

a $\frac{1}{3}$ of 24 is ☐

b $\frac{1}{4}$ of 12 is ☐

c $\frac{1}{6}$ of 12 is ☐

d $\frac{1}{3}$ of ☐ is 12

Comparing and ordering fractions

1 Write a fraction to show how much of each diagram is shaded.
Then write the symbol >, = or < to compare each pair.

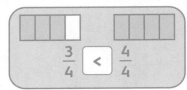

$\frac{3}{4}$ < $\frac{4}{4}$

a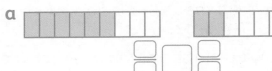

b

c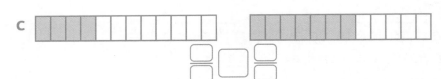

2 Write each set of fractions in order on the number lines.
Look at the example in part **a**.

a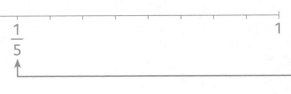

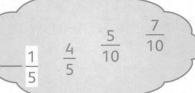

$\frac{1}{5}$ $\frac{4}{5}$ $\frac{5}{10}$ $\frac{7}{10}$

b

$\frac{1}{2}$ $\frac{7}{8}$ $\frac{3}{4}$ $\frac{3}{8}$

3 Write four fractions with **even** denominators in each section of the table.

Less than $\frac{1}{2}$	Between $\frac{1}{2}$ and 1

Adding and subtracting fractions

1 Draw jumps on the number lines to complete each calculation.

a

| 0 | $\frac{1}{8}$ | $\frac{2}{8}$ | $\frac{3}{8}$ | $\frac{4}{8}$ | $\frac{5}{8}$ | $\frac{6}{8}$ | $\frac{7}{8}$ | $\frac{8}{8}$ |

$\frac{3}{8} + \frac{3}{8} = $ _____

b

| 0 | $\frac{1}{10}$ | $\frac{2}{10}$ | $\frac{3}{10}$ | $\frac{4}{10}$ | $\frac{5}{10}$ | $\frac{6}{10}$ | $\frac{7}{10}$ | $\frac{8}{10}$ | $\frac{9}{10}$ | $\frac{10}{10}$ |

$\frac{9}{10} - \frac{5}{10} = $ _____

c

| 0 | $\frac{1}{10}$ | $\frac{2}{10}$ | $\frac{3}{10}$ | $\frac{4}{10}$ | $\frac{5}{10}$ | $\frac{6}{10}$ | $\frac{7}{10}$ | $\frac{8}{10}$ | $\frac{9}{10}$ | $\frac{10}{10}$ | $\frac{15}{10}$ |

$\frac{7}{10} + \frac{6}{10} = $ _____

2 Use two digits each time as numerators, to make each statement true.

| 2 | 7 | 3 | 6 |

a $\dfrac{\square}{8} - \dfrac{\square}{8} > \dfrac{1}{2}$

b $\dfrac{\square}{10} + \dfrac{\square}{10} < 1$

c $\dfrac{\square}{10} + \dfrac{\square}{10} = 1$

d $\dfrac{\square}{6} - \dfrac{\square}{6} < \dfrac{1}{2}$

e $\dfrac{\square}{9} + \dfrac{\square}{9} > 1$

3 Elok's uncle spends his money each week in different ways.

Food	Travel	Cost of running his home
$\dfrac{3}{11}$	$\dfrac{2}{11}$	$\dfrac{4}{11}$

What fraction of his money does he have left to spend on other things?

_____

_____

4 Fill in the missing fractions in these puzzles.
The fraction in the circle above is the sum of the two fractions below.

a

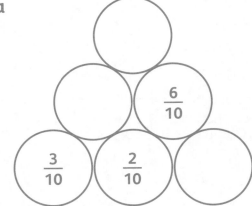

b

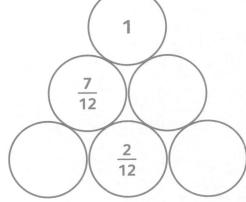

Introducing percentages

1 What percentage of the blocks is shaded in each diagram?
Also write the percentage as a fraction. Do not use denominator 100.

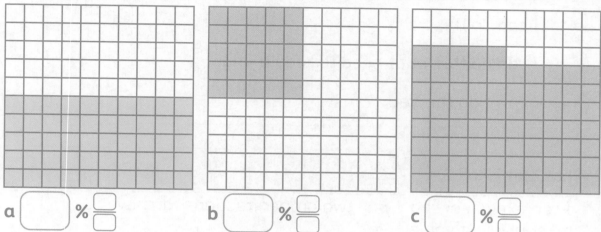

a [ ] % ⬚/⬚   b [ ] % ⬚/⬚   c [ ] % ⬚/⬚

2 Draw three different shapes on the grid. Label them A, B and C.
Shade the shapes to show these percentages:

Shape A, 50 %   Shape B, 25 %   Shape C, 75 %

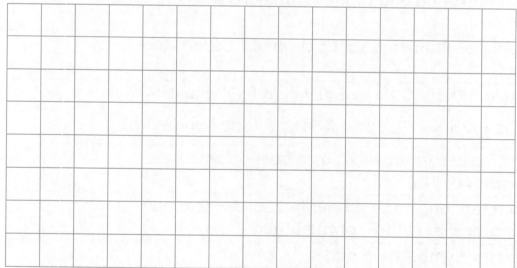

3 There are 100 marbles in a bag.

$\frac{1}{100}$ of the marbles are red. 25 % are yellow.

$\frac{1}{2}$ are green. The rest are blue.

How many marbles are blue? _____

Self-check

See how much you know!

 I can do this.

 I can do this, but I need to keep trying.

 I can't do this yet.

What can I do?			
1 I can explain or show why two proper fractions can have an equivalent value.			
2 I can identify simple fractions with a value or total of one and explain why they equal one.			
3 I can apply understanding of fractions to problems.			
4 I can estimate, add and subtract fractions with the same denominator.			
5 I can apply addition and subtraction of fractions to problems.			
6 I can explain that 100 % is equivalent to 1 whole, 50 % is $\frac{1}{2}$, 25 % is $\frac{1}{4}$, 75 % is $\frac{3}{4}$, and 1 % is $\frac{1}{100}$ (one hundredth).			
7 I can write a percentage (%) as a fraction with a denominator of 100.			
8 I can compare and order proper fractions where one denominator is a multiple of the other.			
9 I can use the symbols =, > and <.			

I need more help with:

_____

_____

_____

Can you remember?

Plot these coordinates. Join them to make a shape. Name the shape.

(2, 3)
(2, 9)
(10, 5)

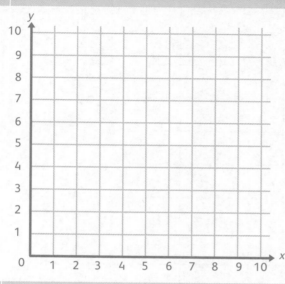

The shape is a:

_____

More coordinates

1 Plan a route through the maze from × to ●.
Your route must go along the grid lines.
Write the coordinates of all the points where your route makes a turn.

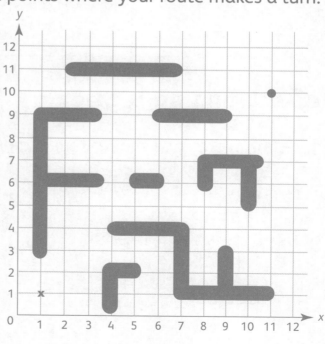

2 Join the dots to make four different squares.
Write the coordinates of the vertices of each square in the table below.

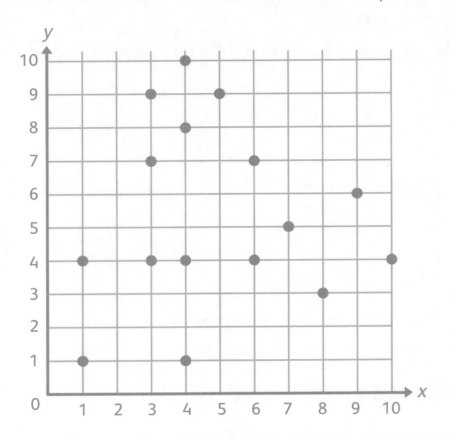

Square 1	Square 2	Square 3	Square 4

Reflections on a grid

1 Complete each shape. Reflect along the dotted lines.

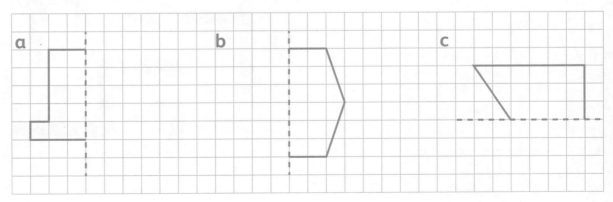

2 Show all the mirror lines.
a Draw four different symmetrical triangles.

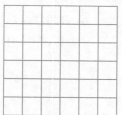 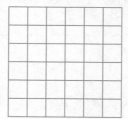  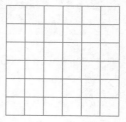

b Draw four different symmetrical polygons.

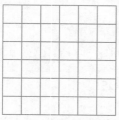 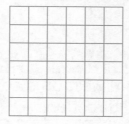 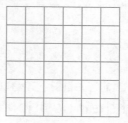 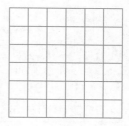

c Draw four different shapes with two lines of symmetry.

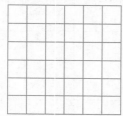 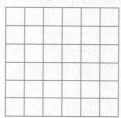  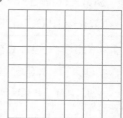

3 Draw a shape above the mirror line. Swap with a partner.
Draw the reflection. Write the coordinates of the new vertices.

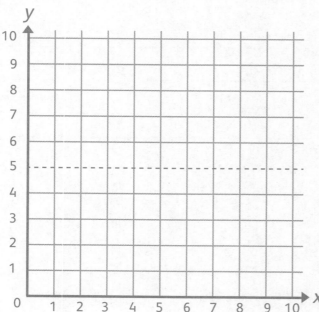

Coordinates:

Unit 18 Angles, position and direction

Self-check

See how much you know!

	I can do this.
	I can do this, but I need to keep trying.
	I can't do this yet.

What can I do?			
1 I can identify north, east, south and west on a compass, and understand their position and direction.			
2 I can identify north-east, south-east, north-west and south-west on a compass, and understand their position and direction.			
3 I can read and find coordinates on a grid, knowing that coordinates are given in the order: *x*-coordinate, *y*-coordinate.			
4 I can use a grid to reflect shapes and patterns.			
5 I can use coordinates when reflecting shapes on a grid.			

I need more help with:

_____

_____

_____